THE GRAPE CURE

Johanna Brandt

www.snowballpublishing.com

info@snowballpublishing.com

For information regarding special discounts for bulk purchases, please contact Snowball Publishing at:
sales@snowballpublishing.com

www.snowballpublishing.com

info@snowballpublishing.com

For information regarding special discounts for bulk purchases, please contact
Snowball Publishing at

sales@snowballpublishing.com

THE GRAPE CURE

JOHANNA BRANDT

———

Printed in the USA

PUBLISHER'S NOTE

This book was first published in the United States under the auspices of Harmony Center, Inc. as a contribution by the author to a suffering humanity and in the hopes of bringing relief to millions of sufferers thru-out the world. It is not to be used to promote the sale of products that might be mentioned herein.

Should doubt exist as to the seriousness of the illness, the services of a qualified physician should be called upon without delay.

CONTENTS

Page

CONTENTS

AUTHOR'S NOTE

In the present desperate need of the world, I am offering this book as my contribution toward the solution of the Cancer problem. It is founded on personal experience and is put forward more as a prevention of Cancer than a cure.

Many of the sufferers from this malady have had organs removed. Havoc has been wrought by the virulent cancer poison. They all have the same story to tell, such as:

"We are now declared to be suffering from inoperable cancer."

To all such afflicted beings—and there are many—I would say that the grape diet is, so far as I know at present, their only hope.

Any questions with regard to such cases should be directed to licensed drugless healers or to the medical men who have studied the Grape Cure.

Before starting on the grape diet, it is absolutely essential to remove any and all prejudice from our minds and hearts. Start with a clean sheet, then the only thing you will have to eradicate is the obstruction in your physical body.

JOHANNA BRANDT,

Pretoria, S. A.

Chapter I.

THE FOURTH OF JULY, 1927

IT was mid-winter when I left my home in the Transvaal to bring the message of the discovery of a remedy for cancer to the United States of America.

Nothing could have been more dreary than the dusty little platform of our provincial town. Something clutched my heart when I looked on the faces of the children who had helped to get mother ready for her strange expedition. When would I see them again? Matters were not improved by the fact that my husband's face was missing. He was away from home on affairs connected with our Church.

It was the 4th of July—the American Day of Independence. This was a mere co-incidence. The date had not been pre-arranged because of its significance but because it fitted in with the lectures I had to deliver in Bloemfontein and Cape Town before sailing for England by the "Windsor Castle."

It was a good omen, I told the children. America was a free country politically, and an independent, powerful, progressive, rich and enlightened nation. But it was not free from disease. I had no doubt whatsoever that this free nation would accept my message, and, accepting it, be blessed with a new emancipation—a wonderful deliverance from disease and premature death. I tried to conjure up visions of the blessed and beautiful state of the world when, through America, a perishing humanity had been saved from suffering and the poverty which so often follows in the wake of disease.

In Cape Town, after one of my lectures, an astrologer who happened to be present, volunteered the information that planetary influences were against my enterprise. I was earnestly advised to cancel my voyage and to return to the Transvaal.

This was discouraging! To hide my deep depression I smiled and said:

"I shall overcome all planetary and other evil influences, by the grace of God!"

* * *

The brooding majesty of Table Mountain enveloped me in a parting benediction.

Disappointment followed in my wake. Every plan was frustrated; my funds ran low and I was so much delayed in England and Europe that it was the end of November before I arrived in New York.

Perhaps some day the story may be

written of how in the end, by the grace of God, every obstacle was overcome.

The first three months in America were difficult indeed. I found to my great disappointment that the Medical Practice Act of the State of New York was tyrannical in the extreme. Much time was lost in constructing a plan by which I could demonstrate the efficacy of the Grape Cure.

As a law-abiding citizen of South Africa, I had no desire to come into conflict with the law of a strange land. There was nothing to do, therefore, but to secure the co-operation of registered medical men and carry out my healing campaign under their protection.

But would it be possible to find medical men who would be willing to supervise test cases under an unknown system of healing?

The time spent in searching for them

was not lost. I visited many people and institutions, presented by letters of introduction, delivered private lectures and worked up many valuable connections. My main activity, however, was writing. The little portable "Corona" typewriter that has accompanied me everywhere since 1916 was nearly worn out with the letters I wrote to the editors of leading newspapers and magazines, the heads of healing movements, the pastors of churches and, last but not least, the most prominent medical men connected with the campaign against cancer.

But these efforts met with no success. The months went by and I did not even get an acknowledgement of the receipt of any of my communications.

Two years before when I was lecturing in Cape Town, I met a fine American woman who was interested in healing and who still had time, on her trip around the world, to help me with my work. We became close friends. Her home in Long Island received me after I landed in New York.

"It is God, who builds the nest of the blind bird."

I still have the latchkey of that home. The refuge is always ready.

Those who have drunk deeply of the cup of homesickness, will understand. But this was no ordinary homesickness. It was not merely a longing for home and loved ones, or a yearning for the "slumbering, sunlit vastness" of South Africa. It was a state of mental and spiritual anguish charged with the unfathomable suffering of all the ages. It was my utter helplessness.

To hold the key to the solution of most of the problems of life and to have it rejected, untried, as worthless—that is to pass through the dark night of the soul. To have the mockery of worldly splendors thrust upon one as a substitute for an ideal; that is the temptation in the wilderness. To offer the gift of deliverance from pain, freely, without money and without price and to see it spurned—that is crucifixion—Calvary.

* * *

THE TURNING OF THE TIDE

Among others, I had a letter of introduction to the Father of Naturopathy in America, Dr. Benedict Lust, and when I placed my difficulties before him he advised me to approach Mr. Bernarr Macfadden, editor of the "Evening Graphic" and the famous magazine "Physical Culture."

Mr. Macfadden received me very kindly. In spite of the fact that I was still withholding the secret of the Grape Cure (until it could be brought forward in such a way that it could never be disputed), he listened attentively to my story and finally invited me to write an account of the discovery for the "Evening Graphic."

What seemed to impress him most was the fact that I was prepared to undergo an exploratory operation to prove my claim, for I have always maintained that the scars of the malignant growth were still present in my body.

This proof of my sincerity touched him and he made a special feature of my case in a full-page article in the "Evening Graphic" of January 21st, 1928.

Chapter II.

MY STORY OF THE DISCOVERY

In the Magazine Section of the New York "Evening Graphic" on January 21st, 1928, my article was published as follows:

By Johanna Brandt

I was born in the heart of South Africa in 1876. Over fifty years ago my forefathers were heavy meat eaters and practically lived on game, as did most South Africans in those days. I do not know whether this has anything to do with the fact that cancer is the greatest scourge of our country, but I think so.

There was a lot of cancer in my father's family and my mother died of cancer in 1916. The doctors tell us that

-11-

the disease is not hereditary. This may be true, but the predisposing causes of cancer in my mother's body may have been present in my own.

It is not unreasonable to assume this. Be that as it may, as long as I can remember I suffered from gastric trouble, bilious attacks and stomach ulcers.

It is cruel, when one is of a highly romantic temperament, to have to turn one's internal organs inside out for public inspection.

Why could it not have been something less prosaic? Heart disease, lung trouble or a delicate throat? But stomach! A reeking fermenting stomach, and a blatantly conspicuous one at that!

After the anguishing spectacle of my mother's martyrdom, I had one shock after another. National, family and other troubles. Life became a ghastly nightmare and through it all I was conscious of a gnawing pain at the left side of my stomach.

Cancer? I was not afraid of it. In my ignorance I thought I had reached the limit of human endurance. I saw in cancer a possible release.

A friend, meeting my husband one day, inquired after my health, and was so much struck by this reply that she repeated it to me:— —

'What must I say about my wife? The hope of death is keeping her alive and the fear of life is nearly killing her.

The hope of death! That was it. But I was puzzled to know how my secret had been discovered.

My plan of action was carefully prearranged. I would allow nothing to be done that could prolong life. If it were really cancer, no medicines would be taken to check the disease. No injections. No drugs to alleviate pain. And, under no circumstances, the application of the surgeon's knife.

At this time a little book was put into my hands, 'The Fasting Cure,' by Upton Sinclair. It thrilled me. A new hope surged through me, the hope of relief from suffering. Here was something that appealed to my common sense. Something constructive—Nature Cure.

The book set the fasting ball rolling in our house, I fasted for seven days. The result was disappointing.

Starting Fasting Class

Nothing daunted, I fasted again and persuaded every one else to fast. In time I set up a fasting business, free of charge. Any one and every one could fast for nothing under my supervision. I became highly experienced and seemed to cure every one, except myself.

The study of one system of healing led

to another. Our home was stacked with the best American books and magazines on the science of spinal adjustment, German water cures, Swiss sun bathing, Russian fruit cures and Oriental works on the science of deep breathing.

A flame had been lighted that nothing could extinguish. It was a pleasure to see our large family of sons and daughters growing tall, strong and athletic. I once overheard the following fragment of a good-humored argument between two small sons:—

'You talk more nonsense in a day than Charlie Chaplin does in a week. Eat more fruit, man. You will feel much better.'

'Fruit! What you want is a jolly good fast.'

We Fletcherized raw carrots and peanuts until our jaws ached. We began the day with spinal exercises and finished it by sleeping outside.

-15-

The whole family joined hands with me in the campaign against disease. Our fortune was spent in building up a system of natural healing so perfect in its simplicity and economy that it would meet the needs of the farming population in the remotest regions of South Africa.

I wrote books and answered thousands of letters, but under it all I knew that my own internal trouble was not responding to Nature Cure.

Nine Years Battle for Life

My battle for life lasted nine years. I fasted myself to a skeleton. I fasted beyond the starvation point, which is a most unusual proceeding, consuming my own live tissues in the effort to destroy the growth. With every fast the growth was unmistakably checked. But it was not destroyed. On the contrary, it seemed to take a new hold on me whenever I broke the fast. *Because I took the wrong foods.*

How Cancer Thrives

I knew exactly what was taking place. I knew that it was wrong to undermine the system by injurious fasting and then to nourish the growth by wrong feeding.

What was I to do? There was no one to advise me, but while experimenting on myself I was learning something new every day.

Among other things, I learned that *cancer thrives on every form of animal food—the more impure, the better.* I suffered from horrible and disgusting cravings for blood—for beef and pork and rich blood-sausages—for stimulating and highly seasoned foods.

The growth was now pushing its way through the diaphragm, toward the heart and left lung. I seemed to see it like a red octopus feeding on the impure blood at the base of the lung. Breathing became difficult. I spat blood occasionally.

One night in August, 1920, I had a terrible attack of vomiting and purging, with excruciating pain. Toward morning I brought up a quantity of half-digested blood.

In Serious Condition

Matters were becoming serious. The thought of the death certificate and possible complications troubled me. I sent for our family physician.

He ordered me to lie still in bed for three months. Under his supervision I fasted twelve days. Plenty of time now to write glowing accounts of the wonders of Nature Cure to distant correspondents.

More than ever I realized the importance of saving my own life in order to convince and try to save others.

It was under this fast that I first

noticed an ominous sign, the presence of digested blood—known in medical circles as "coffee grounds"—in the stools after the use of the enema. Still more disconcerting to find that I no longer put on weight on breaking the fast.

Toward the end of 1920 I seemed to be fasting chronically, four, seven, ten days and finally three weeks in December.

Nothing has been said in this article about the mental aspect of healing. The subject is too big. It forms the most thrilling story of my life, but I must now be content to state that I became superconscious. I had unerring 'hunches' and cultivated a bowing acquaintance with my subliminal self—whatever that may be.

All this fasting brought about a slight improvement and I dragged through 1921 somehow. Then in November I was persuaded by my doctor to go into the General Hospital in Johannesburg for an X-ray examination.

Many plates were taken, and a noted surgeon pronounced his verdict—the stomach was being divided in two by a vicious, fibrous growth. *An immediate operation was recommended as the only means of prolonging my life. This I refused.*

The famous doctor who was operating in the X-ray department was much interested in my experiences and invited me to his house for another X-ray examination if I found myself still in the land of the living after six months.

Encouraged by this mark of sympathy, I fasted three weeks in December, drinking pure water only and lying in the morning sun. When, after six months, I went under the X-ray again, no trace of the growth could be found!

But Pain Remained

I assured the doctor, however, that the pain was still there and told him that I was looking for a food that would answer a threefold purpose, viz.: destroy the growth effectually, eliminate the poison and build new tissue.

The three years that followed were years of great suffering, but I kept on fasting and dieting alternately, and in 1925, after a seven-day fast, *I accidentally discovered a food that had the miraculous effect of healing me completely within six weeks.*

The publication of this discovery will be of more value after the particulars set forth in this article have been proved to be facts.

I therefore call upon the medical council to have an exploratory operation performed. The gravity of the disease can only be estimated by an examination of the extent of the damage done, and then only can the efficacy of the cure be established.

A METHOD THAT MAY CURE CANCER MAY CURE ALMOST ANY OTHER DISEASE. WHAT IS MORE, IT MAY PREVENT CANCER AND ALMOST EVERY OTHER DISEASE.

While I was experimenting on myself, I was often discouraged by the thought that very few people would be able to undergo such rigorous treatment.

But it is the sum total of my experience that I hope to bring before the public. *Fasting for such a long period was unnecessary. The mistake I made need not be made by other patients.* Our system of healing has been greatly modified by the discovery of the food cure. And while the patient is undergoing the cure for his or her own particular complaint, Nature is secretly restoring and rejuvenating every part of the body.

The senses become abnormally acute; dim eyes brighten; faded hair takes on new gloss; the lifeless, hopeless voice becomes vibrant, magnetic, and the complexion clears.

I have seen beautiful sets of teeth, loose in their suppurating sockets, become steady and fixed within a few weeks, the gums free of pyorrhea within a few months.

I have watched our old people getting young and our young people becoming superbly beautiful, and with every new entrancing revelation of the wonders of Nature Cure I have dedicated my life anew to this joyous work of spreading the good news.

The foregoing article created widespread interest. An afflicted nation stirred to the chord of hope that had been struck. I was overwhelmed with correspondence and visits.

This led to unexpected developments.

That Saturday morning was a landmark. I had an informal luncheon party in my hotel to celebrate the publication of an article which I believed would revolutionize healing throughout the world. It was amusing to listen to the outcry of my friends against the proposed operation. They begged me not to consider it, but it was too late to withdraw. It had gone out in the form of a challenge to the medical fraternity and if they accepted it, I would be bound in honor to submit to it.

On the 21st day of January, a stranger called at my hotel—a medical man and a surgeon! The "Evening Graphic" was only a few hours old and already had attracted the attention of a doctor whom I thought had accepted my challenge.

This kindly, enthusiastic Doctor had no designs upon me. The purpose of his visit was to encourage me, and urge me to be steadfast, not to be dissuaded from my plan. Nothing could be finer than such evidence of devotion to a cause. he said.

Afterwards I found out that he was a member of the medical profession of exceptionally high standing, and when letters came pouring in from every part of the United States and Canada, in response to my article, I consulted him.

From other medical men there was no response to my challenge.

Chapter III.

Motto of H.M. is:
"FOR GOD AND COMMON SENSE"

On this, our little planet, the sons and daughters of God groan and travail in unspeakable anguish to bring forth the harmony of the spheres.

Harmony Means Perfection

There is no way of attaining perfection of body, mind and spirit except by applying the Eternal Principle of Harmony to the daily life.

In the hope that the teachings that had saved my own life would be helpful to others, I spent my time in New York in trying to establish a Centre.

The ways of Providence are past finding out. It was a sore trial to be kept waiting for the sympathetic response of the medical fraternity.

One month I waited and then, as no surgeon volunteered to investigate my claims, I formally withdrew my challenge.

Another and better way had in the meantime been found of demonstrating the efficacy of my discovery. If others afflicted with cancer could be cured, then my theory would be proven.

The many heart-breaking appeals for relief could not be ignored. As the laws of the land did not forbid me to tell the story of how I had cured myself, I simply related my experiences and described the procedure I had adopted. People treated by my methods recovered. They in their turn told their relatives and friends—always with the same results.

Correspondents clamored for information about the Grape Cure. At first we sent out typed copies, but when the demand became excessive we had a four-page leaflet printed. Five editions of this were printed. Newspapers in distant states reproduced it and inquiries came pouring in thick and fast. This leaflet, which was distributed free of charge, became famous.

When it became necessary to have a secretary, a woman with great executive ability stepped forward and offered her services. Her rooms were placed at my disposal for the reception of visitors. Our surprise may be imagined when we found that the physician, who had called at my hotel had his office in the same building.

The cancer patients who came to us were referred to him.

Other medical men volunteered to experiment on their cancer patients. No charge was made for the treatment of these test cases.

Chapter IV.

THE FIRST TEST CASES

Our test cases will go down in history.

Those of the patients who knew that they were suffering from cancer were greatly helped by the thought that there was more at stake than their lives. They showed the proper fighting spirit when they understood that the success of a noble cause depended on their efforts and especially on their implicit obedience.

The details of all our test cases have been carefully tabulated for publication. In this volume, however, I am merely able to give an outline of our experiences. When the time comes our physicians will speak for themselves.

Everyone concerned with these test cases in New York passed through a baptism of fire, for, as it happened, the patients were pronounced to be in the final stages of cancer.

I maintain that it is not fair to expect such cases to recover under the Grape Cure or any other treatment.

And yet they have all passed safely through the test!

One young woman had had six operations on the rectum and the base of the spine. I never saw anyone so completely poisoned.

After beginning the grape diet (and she continued it longer than any of the other patients), the pus poured from her. When she began to pass worms, I knew that the terrible ordeal was nearly over. The grapes seem to ferret out the most deep-seated cause of trouble and drive it from the system.

This woman's life will be a perpetual martyrdom. The coccyx had been removed—a physician said she would never be able to sit properly and nothing could relieve the excruciating pain in the spine.

But to look upon that transfigured face fills one with awe. Her eyes shine with an unearthly beauty, her skin is soft and smooth like the petals of a rose. Sanctified by suffering, this woman has emerged from the abyss of premature death to be a witness to the divine healing properties of the grape.

There was a patient in the Bronx—a middle-aged woman, the mother of a large family. When first I visited her, she said she was in the final stages of cancer of the stomach and bowels. Vomiting night and day! When this stage has been reached, it is not considered wise to begin drastic treatment. But that death chamber was charged with so much filial love and passionate anxiety that I had not the heart to refuse my help. Just a few grapes were given at a time. Within twenty-four hours the vomiting stopped. The desperate strain was relieved. But the emaciated victim went steadily down, passing through all the phases of debility and weakness until she actually reached the unconscious stage. The grim struggle for life lasted nearly two months. One healing crisis succeeded another. Finally her legs began to swell.

"This is the end," one of her sons whispered.

"Yes," I replied. "This is the end of the Cure." The poisons had now been collected in a safe place. The family members were instructed to wrap the swollen limbs in grape compresses in order to open the pores. In a day or two—tomorrow, the swellings would probably subside.

"Tomorrow" found a distinct improvement in the condition and soon no trace was left of these dropsical symptoms.

Wonderful to relate, the hard mass in the ascending colon had been disappearing gradually. Nothing was left of that. And the stomach was now so normal and strong that the patient's incessant demand was for "Food."

At this point I must make special mention of the fact that during the most critical period, when the patient was no longer able to take grapes, the pure juice of grapes was administered with a spoon every ten or fifteen minutes. This natural stimulant seemed to tide her over the supreme crisis. I am quite sure that she could not have been saved in any other way. As she grew stronger, she drank grape juice by the pint and gradually other fruits were added to the menu. She is now able to indulge in a delicious variety of raw salads, a dish of sliced tomatoes with olive oil, a ripe banana mashed with sour cream, and her favorite beverage, buttermilk.

Life is full of problems and the great problem in that home at present is the catering department. With the next meal ever looming big in the mind of the patient, ingenuity is taxed to the uttermost to devise something new in the raw food line.

Some of the other cases have passed through experiences no less remarkable. In research work, the most satisfactory experiment may be made with external cancer. While the system is being drained of its poisons under the grape diet, the wounds are kept open with frequent applications of grape poultices, compresses or fomentations. The discharge from these wounds in the early stages of the treatment is horrible in the extreme. As Nature works very thoroughly, the discharge may continue for weeks. The grape apparently eats its way deeper and deeper into the diseased flesh and the wounds do not heal until all the poisons have been eliminated. *Then the healing seems to proceed from within.* No scabs or crusts are formed so long as the wounds are kept moist, but from the glistering bone outwards the process of reconstruction is carried on. Healthy, rosy granulations of new flesh appear and gradually the cavities are filled up.

With such miraculous evidences that a certain cure for cancer may have been found, one is thrilled with a new enthusiasm.

"How Do You Know?"

I have often been distressed by the questions:

"How do you know that you had a cancer?"

This was proven by X-ray examination taken at the General Hospital in Johannesburg. Some of our test cases may likewise be disputed, but *what about external cancers?*

    I am thinking of a woman with cancer of the breast. One breast amputated, the other heavy with suppuration, ready to be removed. More appalling still, pus forming again in the scarred remains of the tissues on which the operation had been performed—these ominous signs striking a chill to the heart.

Then the transformation! A few weeks later during the Grape Cure there was a distinct softening of the swollen breast. No longer overspread by that ugly-looking, dark red hue, it showed a tinge of healthy pink here and there. The wounds on the other side were undergoing an even more remarkable change. There was less pus, and, it was less virulent, less offensive. The patient was exhausted but brave and sweet, she still came to a consulting room to be medically examined.

"Those grape poultices have a miraculous effect," her devoted husband exclaimed.

How are such results brought about? The very simplicity of the diet is a handicap when one reads the directions.

DIRECTIONS

for

THE GRAPE CURE

MEMORANDA

Chapter V.

DIRECTIONS FOR
"THE GRAPE CURE"

(1) *Preparation.* To prepare the system for the change of diet, the better practice is to fast for two or three days, drinking plenty of pure, cold water and taking an enema of a quart of lukewarm water daily with the strained juice of one lemon therein.

By this short fast, complications may be avoided. The stomach is cleared of poisons and fermenting accumulations to a certain extent, and the grape can begin its work more quickly.

The preliminary fast, furthermore, has the advantage of giving the patient a keen relish for the first grape.

(2) *After the fast.* The patient drinks

one or two glasses of pure, cold water the first thing in the morning.

(3) First meal. Half an hour later the patient has his first meal of grapes. Wash them well. (Chew the skins and seeds thoroughly and swallow only a *few* of them as food and roughage.)

(4 *Time*. Starting at 8 a.m. and having a grape meal every two hours till 8 p.m., this would give seven meals daily. This is kept up for a week or two, even a month or two, in chronic cases of long standing. Not longer under any circumstances.

(5) *Variety*. Any good variety may be used—purple, green, red, white or blue. Hothouse grapes are better than none, and the seedless varieties are excellent. The monotony of the diet may be varied by using many varieties. Different varieties contain different elements so it is advisable to use as many kinds as one

can get. Some like them acid, others like them sweet.

The best time is when the grape season is at its height.

(6) *Quantity*. This varies according to the condition, digestion and occupation of the patient. It is well to begin with a small quantity of one, two or three ounces per meal, gradually increasing this to double the quantity. In time about a half pound may safely be taken at a meal. To make this point quite clear, a minimum quantity of one pound should be used daily, while the maximum should not exceed four pounds. Patients taking larger quantities at a meal should allow at least three hours for digestion and should not take all the skins. Invariably, the best results have been effected when grapes have been taken in small quantities.

(7) *Enjoyment*. A loathing for grapes

may indicate the presence of much poison in the system and the need of another short fast. Adding grapes or any other food to such a condition would, therefore, be injurious. The rule in such cases is to abstain from every form of food, drinking an abundance of cold water. Unless patients can eat the grapes with perfect enjoyment, they are better off without them. Skip a few meals. Let Nature regulate this matter. We hear of over-zealous relatives forcing grapes down the throats of unfortunate patients. This is a great mistake. (Always remember that grapes are nourishing and maintain life in the body while the cleansing process is going on). *Loss of strength is due to the presence of poisons in the system.* The patient continues to weaken under the grape diet and under the complete fast, until the poison has been expelled. Then, without a change of diet (and in case of a complete fast, without any food whatsoever), the patient returns to strength and in some cases even puts on weight.

It is a well-known fact with scientists and physiologists that a person can go from 90 to 115 days without any food and live, and that he can go without water 12 days and live.

FOUR STAGES

There are four stages in a complete treatment and these stages must be followed closely; heavy foods must not be eaten until the completion of the four stages.

At the conclusion of the exclusive diet, the patient is in much the same condition as a Typhoid Patient when the fever subsides. Extreme care must be taken to prevent him from eating heavy foods.

First Stage

(a) In every case reactions are different. It is, therefore, impossible to say beforehand how long it will be necessary to use grapes only. But this may be stated definitely—the cleansing of the elimentary canal takes time, and until this has been accomplished, the real relief does not begin. It is safe to say that the first seven to ten days on grapes only would be required to clear the stomach and bowels of their ancient accumulations. And it is during this period that distressing symptoms often appear. Nature works thoroughly. She does not build on a rotten foundation. The purification of every part of the body must be complete before new tissue can be built.

I think this is the only explanation of the excessive loss of weight under the grape diet.

This question is of so much importance that we refer to it in detail under the treatment of cancer elsewhere.

If we could remove every trace of fear from the mind of the patient, the correct procedure would be to continue the exclusive grape diet *until he stops losing weight*. By watching the symptoms—the temperature, the excretions, eruptions, etc., we know when the work of purification is complete. When this point has been reached—and it may last from two weeks to two months—it is advisable to go on to the

Second Stage

(b) The gradual introduction of other fresh fruits, tomatoes and sour milk or cottage cheese.

We do not expect anyone to live on grapes forever. The grape contains many of the most valuable elements necessary for life, but it does not contain *every-*

thing. To live on grapes indefinitely would be to rob the system of some of the elements essential to life. When we are sure, therefore, that the grape has done its work by breaking up the unhealthy tissue and purifying the blood, the careful introduction of other body-building foods is the next step.

Grapes still form the main food and are always taken as the first meal in the morning and at 8 p.m. But now, during the day, some other fresh fruit may be used instead of grapes. An endless variety presents itself—a slice of melon, an orange, a grapefruit, an apple, a luscious pear, the scarlet strawberry, the golden apricot—one fruit more appetizing than the other.

Let the patient choose.

Only one kind of fruit to be taken at a meal but something different every day.

After a few days a glass of sour milk or buttermilk, yogurt, or cottage cheese may be taken instead of grapes for supper. Patients who dislike milk should take a ripe, finely-mashed banana, or some other nourishing fruit.

After a week or ten days, every other meal may consist of different varieties of fruit, or sour milk, taking them, for example, in the following order:

8:00 A.M. Grapes.

10:00 A.M. Pear, banana or peaches.

12:00 Noon Grapes.

2:00 P.M. Sour milk, buttermilk, or cottage cheese.

4:00 P.M. Grapes.

6:00 P.M. Orange, grapefruit, plums or apricots.

8:00 P.M. Grapes.

At this point some patients crave for something savory. The sweet fruits begin to pall. There may even be a positive aversion to grapes, in which case they should be omitted altogether and the other foods taken every three hours.

One or two sliced tomatoes with pure olive oil and a little lemon juice may safely be included in this diet. The tomato is more of a fruit than a vegetable, containing many valuable properties, and it forms an indispensable part of the diet in the second stage of the treatment.

Third Stage

(c) *The raw diet.* This includes every food that can be eaten uncooked—raw vegetables, salads, fruits, nuts, raisins, dates, figs and other dried fruits, butter, cottage cheese, sour milk, yogurt and buttermilk, honey and olive oil.

Begin the day as usual with cold water and grapes or some other fruit for breakfast, but instead of sour milk or fruit for lunch, have a substantial salad of raw vegetables. Reduce the number of meals, as raw vegetables require longer to digest.

It is surprising to some people to find that nearly all the vegetables can be used raw—young green peas and string beans, celery tomatoes, cucumbers,) lettuce, sprigs of cauliflowers, squash, shredded cabbage leaves, grated carrots, turnips, beets and parsnips, finely chopped onion and spinach.

After the light fruit diet, it is wise not to start out too soon with a large variety of vegetables. Choose two or three of the above-named as a foundation for your salad and mix them with lemon juice and olive oil. Try different varieties the following day and watch the combinations of flavors. Salad-making is a supreme art.

Above all things, this noonday meal should be made palatable. Patients who have been used to animal food crave for something stimulating. There can be no objection to adding one or two savory ingredients to this salad—some finely-chopped nuts, grated cheese, sour cream, or a good homemade mayonnaise made of eggs, lemon juice and olive oil. In some cases a finely-chopped hard boiled egg may be included in the salad.

Time to Digest

Give this meal more time to digest than is required for raw fruits, especially if nuts, dates, raisins or other dried fruits have been added to it.

The supper should consist of sour milk or fruit, or a highly nourishing and digestible dish may be made of ripe bananas mashed, with sour cream.

The Raw Diet

Sufficient stress cannot be laid upon importance of the raw diet. If we could only educate the people to this fact it would help to eradicate disease.

The raw foods digest more easily than the cooked and pass through the system far more rapidly. The result is that they have no time to decompose in the alimentary canal. There is no undue fermentation and no fear of toxic poisoning.

Therefore patients are strongly advised to abstain from every form of cooked food during the full period of treatment.

Thus far the course then consists of the three stages as outlined above and if followed the highest results are obtained.

When it is difficult to convince people that they derive more nourishment from uncooked foods, we reluctantly consent to the introduction of one cooked meal a day, but do not recommend it.

Fourth Stage
(d) *The mixed diet.*

With this innovation, there is sometimes a recurrence of the old trouble, and the patient, sadder and wiser for the experience, is glad to go back to the raw diet. But if the disease has not been very deep-seated and the cure is complete, the following regimen is recommended:—

Three Meals a Day

(1) A fruit breakfast, one kind only.

(2) A cooked dinner.

(3) A salad supper.

For breakfast eat plentifully of any of the juicy fruits that may be in season. Make a strict habit of this and observe it for the rest of your life if you want to be healthy.

The No Breakfast Plan

does not apply to fruit at all. It was and is, a splendid rule for people who have been systematically overeating, and especially those who are in the habit of indulging in heavy dinners and late suppers. But when the supper is taken not later than 7 P.M. and consists of raw salad or fruit, the stomach of one who has been on a proper grape diet is free from acidity and accumulations.

In such cases, the fruit breakfast is better than the fast, in that it supplies the body with cleansing and building material.

One can, moreover, do a hard morning's work on a fruit breakfast.

Not a Cookery Book

This book on the Grape Cure is not a cookery book.

It would, however, be incomplete without a few hints on cooking for the benefit of the reader who has followed these pages.

As the cooked foods are the acid forming foods, no one who is troubled with acidity should have them.

Raw fruits and vegetables never *cause* acidity—on the contrary, they *neutralize* the acids by which the system has been poisoned. The first results of a raw diet are often very distressing on that account. The patient *seems* to become hyperacid and this condition lasts until all the poisons have been worked out.

Another thing to remember is that cooked foods take much longer time to digest than raw. No more food should be taken within five or six hours after the cooked meal. No strenuous work should be done and it is especially recommended to refrain from every form of brain work immediately following such a meal.

Keep the cooked and uncooked foods apart.

In the process of digestion, Nature always disposes of the most digestible foods first.

If you have made the mistake of mixing raw foods with the cooked, the raw will digest first while the rest of the conglomeration will ferment.

The Cooked Meal

A dry meal. No soups, no liquids of any kind.

No raw salads. No fruit either fresh or cooked.

The main foods to be *steamed vegetables*. Begin with one kind at a time after the Grape Cure. If the results are good, take two or three varieties at a meal.

Not more than one kind of starch. This may consist of any of the cereals, such as oatmeal, wheatena, brown rice, potatoes or whole wheat bread and unsalted butter. Prof. Arnold Ehret, originator of the Mucusless Diet Healing System and recognized authority on raw food objects to even one starch.

Enjoy this meal. If you are not a vegetarian, indulge in a piece of baked, broiled or steamed fish occasionally, with a baked potato and butter. Or this meal may consist of a dish of stewed tomatoes, or any of the green vegetables steamed and baked. An infinite variety of savory dishes may be made by mashing one of the green vegetables with steamed potatoes, mixing with egg, covering with bread crumbs and pats of butter and baking this to a rich brown in the oven. Left-overs of cauliflower, carrots, cabbage, parsnips, steamed lettuce, spinach, baked onions, etc., lend themselves especially to this form of cooking.

Watch the effects of the cooked meal and with the first sign of discomfort, return to the raw diet.

The Seven Doctors of Nature

There are seven Doctors of Nature:

(1) Fasting.
(2) Air.
(3) Water.
(4) Sunlight.
(5) Exercise.
(6) Food.
(7) Mind.

We put Mind at the end because it is the most important.

Mind operates through magnetism. In order, therefore, to contact the forces of Mind, we purify and build up the magnetism and this is best done by fasting, deep breathing, water treatment, sun bathing, spinal adjustment and exercise, the Grape Cure and diet of raw fruits and vegetables.

This is common sense healing or Harmony.

Even the Grape Cure, if not followed scientifically, has it limitations. Take my own case, for instance. Fasting could not save me while I continued taking cooked foods after each fast. The Grape Cure would not have had permanent results in my case if I had not followed the instructions stated herein to get the best results.

We live in the age of invention and discovery. When I stumbled onto the Grape Cure in 1925, I could give no explanation of the wonders it had performed in my body.

It was, therefore, with keen satisfaction that I received the following scientific statement from Dr. La Forest Potter, of New York City:—

"Proteids are the great body builders. If they contain toxins, they become body destroyers. Because most of our conventional foods, medicines and serums are contaminated by these toxins, science has been searching for years for a non-toxic proteid which will not disturb the colloidal integrity of the cells.

The grape diet seems to be in line with this great world search."

Chapter VI.

REGARDING CANCER

It is my belief that the body of the cancer patient contains the most virulent poison and that cancer is the death and disintegration of a given part of a living body.

Under the grape diet, that decomposition is arrested—checked. But the danger is by no means over. The poisons have now been thrown into the blood-stream and carried to every part of the body. Everything must be done to expel them and expel them quickly.

The Enema

Since the alimentary canal is the main avenue of excretion, the bowels must be attended to first. We recommend the daily use of the enema *unless the grapes act as a laxative,* In some cases they cause constipation until the system has learned to use the skins of the grapes as roughage. Such patients should take an enema of one quart of lukewarm water once or even twice a day, until there is a natural movement of the bowels.

This rule applies to all persons who are undergoing a grape diet.

Why the Skins?

The skins, not alone of grapes, but of many other fruits, such as apples, pears, contain immensely valuable elements. To throw away the peels would be to deprive the system of the very substances required to build a new and healthy body.

But the skins also form the bulk and roughage which are needed to promote the peristalic action of the stomach and bowels.

-64-

Until the system has learned to utilize the grape it is advisable to be careful with the seeds and hulls. A normal digestion suffers no inconvenience when the whole grape is used, on the contrary, it is benefitted by the valuable properties contained in the seeds and skins apart from the bulk and roughage they provide. But if you have been in the habit of discarding them, they may at first accumulate in the digestive tract and cause constipation. I therefore advise all who are experimenting with the Grape Diet for the first time to begin with the juice and pulp; later a few of the skins. Chew them all well in order to extract their essences, but swallow only a few until you are sure that your digestion is able to take care of them. The same applies to the seeds, but do not chew the seeds.

But in all cases use the enema if Nature fails to do her work.

Do not depend on faith for this. The

development of faith takes time and there is no time to lose—if it really is a cancer. Throw the vile poisons out. Employ every material means of ridding the system of its gross impurities and you will be surprised to find that in doing this you are developing the more spiritual powers of mind and soul.

Obedience to the dictates of common sense—harmony—has its own reward.

The Course of the Disease

Under the Grape Diet, it should run its full course within a month or six weeks. The patient loses weight to an extent that would be alarming if he did not understand the principle of the Cure. While on the first stage of Grape Diet, nothing should be administered to make him gain weight—no food of any kind except the grape. In advanced cases it is sometimes by reducing him to a virtual skeleton that the disease may be overcome. When in severe cases he has reached this point, when he has been brought almost to skin and bone, there is nothing left for the cancer to live on and it usually disappears spontaneously. There are many kinds of cancer and the patients are all at different stages of the disease. It is therefore impossible to say beforehand how long it will take to arrive at the turning-point.

No Cause for Anxiety

When the inorganic poisons have been effectually eliminated and the patient seems to have reached the last stage of weakness, there is usually a sudden and marked change for the better. He may fall into a refreshing sleep and wake up feeling "fine," feeling strangely invigorated—and usually his first demand is for "food." THE GREATEST CARE MUST NOW BE EXERCISED. For a day or two he should live on pure grape juice (homemade, or unsweetened commercial)—half a pint sipped slowly every two hours, and then gradually the other fruits may be introduced.

The critical time is during that period of exhaustion just before the tide turns. It is so natural to think that the patient is dying for want of nourishment. This is not the case at all. If he has been kept faithfully on grapes, he has been fed with the finest and richest food on earth—a food that will sustain a healthy, active man for months.

"Feeding the patient to keep up his strength" is the surest way of killing him.

No, your cancer patient is not dying of starvation. And he is not getting weak because he is eatinig grapes only. But the cancer is dying. Nature is destroying it. The vital centers—heart, lungs and brain—are being nourished to the last moment by the grape. To change the diet at this moment is not advisable. Remember this when you are tempted to administer other foods and stimulants. His only chance may be the grape. You know the verdict that has been passed—inoperable cancer, no hope, nothing known to science that can save this case. Well, then the grape is the only hope. Advanced cases have apparently been cured. Until you hear of something better than grapes, do not let the harrowing sight of the patient's weakness and emaciation tempt you to offer other foods. That might be robbing him of his only chance.

Give the grape diet a fair chance for a few weeks, *not longer,* if you are unable to get reliable advice. Then try some of the other juicy fruits.

I have seen patients reach the unconscious stage and then recover.

Fortunately, the patients themselves often have a conviction that grapes and grapes only will save them.

To try them, I occasionally suggest some other tempting food and I have been impressed by the emphatic reply: "Not for a hundred thousand dollars!" or

"I would rather die on grapes than live as I have been living on other foods."

All this, however, applies only to the first few weeks of the exclusive grape diet.

The danger of staying on grapes too long is that the patient may in time not be able to take any other food.

* * *

Report Your Progress to Us

In the name of science and humanity we invite cancer and other patients to report the result of the Grape Cure to us. It is most essential to collect extensive data for publication to encourage others.

Must the Cancer Patient Know?

One of the stumbling-blocks to be overcome in the treatment of cancer patients is secrecy. Every device is employed by doctors and relatives to hide their real condition from them. When the principle of the Harmony is better understood, the fear of cancer will be removed and it will   not be necessary to deceive and mislead cancer patients. After all, they are not really blind to their condition and I know by experience that there is nothing more painful than suspense and uncertainty. The secret fear, the haunting suspicion of being deceived are far more trying and harmful than the realization of danger. It is not reasonable to expect the patient to be faithful to the prescribed rules without a full understanding of the gravity of his condition.

Harmony

On the principle of Harmony we do not put inorganic poisons into the system of a patient who is trying to expel poisons. That would be like pouring oil on a fire with one hand and water with the other, in an effort to extinguish it.

The construction of the human body is marvelously complex. The millions of pores on the surface of the skin have many functions. They not only excrete waste products but they have the power of absorbing material essences. The care of the skin is of immense importance in the treatment of cancer, for the pores are accessory organs of breathing. That is to say, instead of having only one pair of lungs, we possess many millions, so that in the case of internal cancer, when the breathing is heavily restricted, we have to depend largely on the free action of the pores for oxygen.

The Use of the Flesh Brush

The skin of a healthy person casts off the dead cells without artificial means, but in a diseased body the pores are clogged and choked with decayed matter. The entire body of the patient should be brushed morning and evening with a dry flesh brush.

These flesh brushes may be procured at almost any drug store.

But it is not enough to get the poisons out, you must see to it that no new poison gets in. There are more ways of taking poison than by way of the mouth.

Trunk packs, water compresses and poultices, on the other hand, are highly commendable, and in extreme cases these should be saturated with diluted grape juice. A great many patients can now testify to the magical effects of these compresses—many lingering sufferers in the future will experience undreamed of relief through these methods.

Chapter VII

SUPPLEMENTARY INFORMATION

It has come to my attention that many patients overdo the exclusive grape diet. They seem to think they must continue the grape diet until the growth has disappeared completely.

Experience teaches that scars caused by a malignant growth remain in the tissues long after the Grape Diet has done its work. Time only will show whether they will ever be entirely eradicated. The same is true of any injury done to the body, through burns, cuts, fractures, etc. When the grape has purified the blood, the general condition of the patient steadily improves in spite of the presence of lumps, scars, or other evidences of the injury done by the growth.

In my own case, while the poisons of the cancer have been eliminated and the cancer has disappeared, physical examinations by medical men disclose that there are numerous adhesions as a result of the malignant growth. One doctor advances the opinion that it will take at least seven years for these adhesions to be broken up. It is useless, therefore, to continue the grape diet in the hope of completely eradicating the growth within a few weeks, or even a few months.

This treatment is slow, requiring patience and perseverance, but the patient is improving all the time, frequently able to go about his daily duties. One cannot expect to rid himself within a few weeks of poisons which he has been *storing up during his entire life.*

Many patients who are employed and do not wish to become too greatly weakened take the exclusive grape diet for, say, two or three weeks; then they go on to the second and third stages for equal periods. If at the end of that time they feel that the poisons have not all been eliminated, they repeat the treatment. In that way they are able to continue their employment.

The condition of each case should be watched carefully and judgment exercised. It is obviously impossible to give a set of rules which can be followed in each case.

The grape contains many of the elements necessary to sustain life and health, but it does not contain everything. To continue the grape diet beyond a reasonable time would therefore be to deprive the system of the nourishment necessary for the maintenance of the body.

Quantity

It has been observed that many patients are eating too many grapes. Two pounds a day, or if the patient is active and out of doors, three pounds, is usually enough. If the patient is not hungry, it is not necessary for him to force himself to eat. Seven meals is not compulsory, nor is it necessary.

We have observed that the best results are obtained where the patients are given small quantities of grapes. Some patients are forced into eating too many grapes by anxious relatives. Sometimes one's loved ones are one's worst enemies at this stage of the diet. For that reason it is well for the patient to go to a sanatorium if it is possible. Even though the patient should lose considerable weight, it does not mean that he is in danger of starving. As mentioned, grapes contain most of the food elements necessary to sustain life; many have been known to live many months on grapes alone, but this is not deemed advisable.

Amount of Water to be Taken

We have received many inquiries regarding the amount of water the patient should drink. The patient is permitted to drink as much water as he feels inclined to between the grape meals. Generally, sufficient water is supplied by the juice of the grape. In the early stages of the diet the patient often becomes very thirsty. Nature calls for an abundant supply of water to flush the system. After the poisons have been eliminated, this craving ceases. Too much drinking often tends to overwork the kidneys.

Grape Poultices

In case of external growths, where

there is an open sore, grape poultices have been found to be effective. The grape poultice is made by crushing grapes, spreading between layers of cheese cloth or muslin, and placing over the effected part, covering the whole with a dry cloth.

Grape Compress

Where whole grapes cannot be procured, a compress may be used. Soft muslin or cheese cloth is dipped in grape juice diluted about two-thirds.

Both poultice and compress should be renewed frequently, as they absorb much of the poison.

The purpose of the compress is to keep external cancers and other wounds open and soft so that the poisons can be easily expelled. This is important.

Bowel Movement

Distressing symptoms occur during the Grape Diet, through the poisons which have been stirred up by the action of the grape and thrown into the blood stream. These symptoms may be aggravated in cases in which there is very poor elimination. Sufficient stress cannot be laid upon the importance of keeping the bowels free by using enemas.

Many patients complain of becoming constipated under the grape diet. The reader is referred to page 77 of this book regarding the consumption of grape skins.

As laxatives are not advisable, a teaspoonful of olive oil occasionally is recommended. This should be taken just before the grape meal. In extreme cases, a small quantity of olive oil may be injected directly into the rectum by means of a hard rubber syringe.

There are however, several herb laxatives which are effective and do not conflict with the grape treatment.

Formation of Gas

Another development of which many complain in the Grape Cure is the formation of gas. When this symptom appears, it is well to stop the consumption of the skins and seeds altogether for a while. If this does not relieve the condition, one of the best reliefs is by the high colonic irrigation in the knee-chest position. If the patient is not strong enough to take this treatment, the cold trunk pack will usually help to dispel the gas.

Soreness of Mouth

When the grape diet makes the mouth sore and raw, it may be because the flesh is diseased. Grapes do not have this effect on healthy tissue. When the body has been cleansed of its poisons, the soreness disappears.

Other Symptoms

Sometimes, after continuing on the grape diet for several weeks, the feces become quite black. This is deemed a temporary symptom and no cause for any uneasiness.

Acute and Chronic

Acute Pains: As heat promotes inflammation and congestion, cold applications externally are recommended to relieve the patient's pain. Cheese-cloth wrung out of lukewarm water and folded several thicknesses may be placed over the affected part and covered with an ice bag. Renew the cloth frequently. This has afforded relief in cases of appendicitis, piles, acute liver attacks, gall stones, kidney stones, inflammation of throat, etc.

Chronic Cases: When the patient is below par, depleted, or suffering from low blood pressure, the pain may be relieved by hot applications. A cheese-cloth can be wrung out of lukewarm water as mentioned before, but instead of an ice bag, a hot water bottle placed over the cloth.

It is the moist heat that relieves pain. The hot water bag is seldom applied to the dry skin. The moisture opens the pores and enables the impurities to come out. The cloth, therefore, should be kept very clean and renewed at intervals.

Caution

Numerous instances have been reported of patients who have been misinformed as to the duration of the Grape

Cure and the quantity to be consumed. This is regrettable because it detracts from the value of this wonderful discovery. The patient should be urged to adhere closely to the instructions given in this book, all these details having been carefully worked out.

The writer does not want to be held responsible for failures due to wrong advice by persons who do not know anything about the grape diet.

The process of the grape when eaten exclusively tends to cleanse the intestinal tract and dissolve poisons which may have settled in any part of the body It is not advisable to use medicines while on the grape diet.

Why Fast?

When preparations containing iron or other inorganic metals have been injected into the blood in the form of serums, the acids of the grape have corrosive tendency. This danger is mitigated by prolonged fasting before beginning the Grape Cure.

Other Diseases

It is impossible in a work of this scope to enumerate all the diseases which have been reported successfully treated by this method. A few about which special inquiries have come in are mentioned below.

ARTHRITIS: It stands to reason that when the bones have become ossified through arthritis and rheumatism, the grape diet will not loosen them up, except when used in conjuction with other methods of natural healing, such as manipulations, gentle exercise and fomentation. This should not be done without the supervision of one trained along these lines.

DIABETES: This method has been particularly successful with diabetes. The grape sugar is believed to be an organic solvent which neutralizes the sugar deposits in the blood.

When taking the exclusive grape diet for Diabetes, persons who have been taking Insulin should cut down gradually on the intake of Insulin until it becomes zero. You cannot mix it with the grape diet and get satisfactory results.

GALL STONES: Gall stones have been reported dissolved while the patients were under treatment for more serious diseases.

CATARACT: The same may be said of cataract of the eye.

DIET: As we believe many diseases are caused by wrong combinations of food, we advise our readers to study the second and third stages carefully for information on this important subject. Many valuable books are being published in our day on the benefits of the raw vegetable and fruit diet.

ULCERATED STOMACH

This must not be treated lightly. Too often gastric ulcers are neglected. Before the diseases becomes serious, it is recommended that the sufferer undergo the grape treatment. Skins, and seeds should be omitted.

TUBERCULOSIS: So far in this country only a few cases of tuberculosis have been treated, but it is specifically mentioned in reference books published in Central Europe as a disease which readily yields to the grape diet.

SYPHILIS: The writer has been informed that several hundred years ago, when the royal houses of Europe had become corrupt and tainted with this dread disease, they resorted to the grape diet.

In connection with the grape diet we have used the following with amazing results.

For open sores—Grape poultice is made by crushing grapes, spreading between layers of cheese cloth or muslin, and placing over affected part, covering whole with a dry cloth. If grapes are not available, a compress may be used. Soft muslin dipped in grape juice diluted about 2/3. Poultice or compress should be renewed frequently as they absorb much poison.

Rectal Cancer—Treat with diluted grape juice enemas (1/3 grape juice).

Nasal Catarrh, Sinus trouble, etc.— Nasal douches 6 or 8 times a day with diluted grape juice.

Cancer of throat or gullet—Gargle with diluted grape juice.

Cancer of the womb — Regular douches, luke warm water (1/3 grape juice).

When effects of grapes become too drastic—Judicious fasting — FAST — and fast again—just pure cold water. Nothing can be done without the fast.

If weakness develops it is the result of poisons circulating thru the blood stream, rather than a lack of food. This is also true of dizziness - increased pains in the various joints of the body and even vomitting.

Chapter VIII

GRAPE JUICE

It seems too good to be true but it is a fact that something has at last been found that effectually solves the problem of what to do when fresh grapes are not procurable.

We have been experimenting with unsweetened, unfermented, bottled grape juice to take the place of whole grapes during the winter months. So far, the results have been most gratifying.

Even during the period when whole grapes are abundant, there are times when the patient becomes tired of grapes;

or unable to chew them. It therefore gave me double satisfaction to learn that grape juice may take the place of whole grapes.

In critical cases, the patient should fast on water only for a few days using the enema daily as recommended and then he should live exclusively on grape juice for another short period.

A glassful (one-half-pint) half diluted is usually given at a meal but more may be taken if patient desires.

It has been found that the patient can get along almost as well on grape juice alone as on whole grapes, although grapes are preferable when they can be procured. The stomach is accustomed to bulk and is more likely to feel the pangs of hunger when only juice is taken.

Raisins have been taken for some of the periods, to supply bulk, with good results. For instance a glass of grape juice on arising, two hours later a cupful, more or less, of raisins; and either grape juice or raisins at two hour intervals for the remainder of the day. They should not be taken at the same meal. The raisins may be eaten dry, or they may be soaked in ordinary cold water for several hours, and the raisin and raisin water taken for one meal. Any brand of raisins that have NOT been sulphured may be taken. When raisins have been sulphured that fact is printed on the box.

In case where neither whole grapes nor unsweetened juice could be procured, patients have taken only raisins and raisin water instead of grape and grape juice. The raisins should be taken at two hour intervals, the same as grapes. If the soaked raisins are too sweet, a little lemon juice may be squeezed into them,

An analysis of grape juice shows that the grape loses none of its healing properties during the process of double sterilization at a temperature of 90°.

In America there are a number of good brands of unsweetened, unadulterated grape juice.

A bad case of cancer of the tongue has been reported successfully treated, (after a fast of ten days on water) by the administration of a tablespoonful of grape juice every half hour.

It is the only nourishment prescribed in cases of congestion in or adjoining the intestinal tract. After a period of fasting, during which the system is cleansed and prepared for the change of diet, grape juice seems to act as a powerful solvent. At the same time the strength of the patient is maintained by its nourishing properties.

Grapes apparently dissolve the hardened mucus adhering to the walls of the stomach and intestines. In several cases, where the fecal matter was observed there was an accumulation of stringy, slimy mucus which looked like worms. Sometimes this occured in long strings, sometimes in clumps resembling balls of string, or sometimes like black and green marbles. The juice appeared to dislodge this accumulation from the walls of the stomach and intestines and act as a flush, carrying it to the rectum. It is therefore considered very important to remove this accumulation by daily enemas, preferably high enemas. When taking only liquids, whether water or fruit juices, daily enemas are deemed necessary, otherwise poisons which have been carried to the intestines by these liquids are likely to be reabsorbed into the blood.

One case was that of a woman who was said to be suffering from double lobar pneumonia, leakage of the heart, bleeding of the kidneys, and other complications. She was told that she could live but a few weeks. This woman was a patient in a New York hospital. A European specialist who happened to be in New York was called in. He was from Austria and had seen the effects of the grape cure in the Tyrol. He ordered her to be fed on unsweetened grape juice, one spoonful at a time, gradually increasing the quantity and the intervals between until the required results were obtained. The physician said to her: "Remember, madam, this is the same as a blood transfusion. Grape juice is a blood maker." She immediately began to gain strength and later claimed to have no kidney trouble and that her heart was sound.

Two years later, while intestinal "flu" was an epidemic and was particularly fatal, this same woman contracted it.

She went to bed, sent out for some grape juice and took it for two weeks. She recovered, and furthermore says she thinks she has finally rid her system of the "pneumonia bug" for all time.

Another case was that of a woman of over forty years. She was suffering from a growth over the instep which had every symptom of malignancy. She was overweight when she *started* on her first diet. During the twenty years previous to her first grape diet she had at times shown symptoms of goitre, defective eyesight, acidosis, pain in the teeth, and pain in the feet, which developed finally into the growth. The first time she took the diet she was on it for ten weeks. After an interval of a year she took another diet of eight weeks; and the following year, a diet of sixteen weeks. It was after the last diet that the discharge of mucus, as outlined above, was noted.

Between the second and third diets, the stomach was extremely sensitive to the touch but after the discharge of mucus the sensitiveness ceased.

It was noticed that during each diet pain was perceptible in different parts of the body, especially in the stomach and intestines. This pain shifted from hour to hour. During the first few days on water or grape juice, swellings appeared in various parts of the body, sometimes very marked but which gradually disappeared. Her general health improved continuously during the time she was taking the various diets.

A pleasing drink to take at night is made by pouring a glass half full with unsweetened Grape Juice and fill up with hot water.

Chapter IX

THE SECRET OF THE SUCCESS OF THE GRAPE DIET

" Proteids are the great body builders."

This, then, partially explains why new tissue is built with such extraordinary rapidity on an exclusive grape diet.

But science has not yet discovered what elements in the grape break down malignant growths. Some fine essence, which has hitherto escaped observation must be present in this "Queen of Fruits."

Perhaps this elusive substance will be found in the pure unsweetened juice of the grape.

We are only on the threshold of our discoveries and experiments, learning something new every day. With great success the pure juice of the grape is used for cleansing the throat, ears, nose and mouth, applied externally on wounds in the form of poultices and compresses diluted with water, and introduced per rectum as food.

But let us see what can be achieved by the grape diet only, without these supplementary methods.

The Mono Diet
(One food only)

I believe it was the *exclusive grape* diet that saved my life in the end.

After the nine years' battle with death, I discovered almost accidentally that fresh grapes, *when taken alone,* answered the three requirements, of dissolving eliminating and building.

Like everyone else, I had been eating grapes for years. I grew up among the vineyards of South Africa and the finest grapes of the world were to be found on

our table at "Harmony" in Pretoria.
We ate them with other foods; that
was a mistake.

The stomach is Nature's own laboratory. Put the right combinations of food
into it and the result is the fabrication
of every essence necessary for life and
health.

But the stomach is also a still. At the
temperature of blood-heat the process of
digestion is carried on. The manufacture
of natural body alcohol takes place. This
seems to be indispensable to the well-
being of the body. But an excess of alcohol causes a poisoning of the system.
Toxemia or auto-intoxication (self-poi-
soning) occurs. This is true when foods
are mixed in the stomach indiscrimi-
nately, and especially so when to this
mixture is added the grape. Taken
into an impure stomach, it becomes an
enemy.

On the other hand, when by fasting the system has been prepared for the change of diet, the grape becomes our greatest benefactor, our saviour from the ills of the flesh.

When there is an ulcerated condition in the stomach and bowels, seeds should not be swallowed.

The grape is, as far as I know, the most powerful nature solvent of some chemical deposits, and at the same time the most drastic eliminator. Because of its extraordinary properties, the avenues of excretion become superbly active under a proper grape diet.

The first results of the diet are therefore different in every case. According to the condition of the patient, the first effects may be distressing or instantly beneficial. A healthy person may go on the exclusive grape diet and suffer no inconvenience, lose no weight and be able to carry on his usual work without any loss of strength. Not so the sick. In an unhealthy body the complications arising from the diet may be in exact accord with the gravity of the disease.

This is the most perfect form of diagnosis and the most natural. It reminds me of the primitive test of the temperature of the baby's bath—when the baby turns red the water is too hot, when the baby gets blue the water is too cold. Poor baby! And one sometimes feels inclined to say, "poor patient!" when under the Grape Cure, slumbering evils and latent diseases begin to manifest themselves. A seemingly healthy person may set out gaily on a grape diet, merely to reduce weight, for instance, and at the end of a few weeks he may be quite a sorry spectacle. Some deep-seated trouble has been ferreted out by the grape, driven out into the open, and the wise thing to do in such cases is to continue the four stages of the diet until every trace of disorder has disappeared.

"Do I Need the Grape Diet?"

This is a question put most frequently. No one can answer it except yourself and that is best done by starting right out on the diet. It cannot possibly harm you. Try it for a week or two and at the end of that time you will probably know more about your condition than you ever did before.

The average person has been taught to regard every symptom of disease as an evil to be suppressed immediately. I believe nothing can be further from the truth. The disease is evil certainly but *the symptoms of disease are curative processes*—not to be suppressed.

This we find to be true under the fast and under every natural system of healing but particularly so under the Grape diet.

Abnormal growths, cancers, tumors, ulcers, abcesses and fibrous masses *seem to be dissolved* by the powerful chemical agent in the grape. Diseased tissues and fatty degenerations, every form of morbid matter, is apparently broken up into minute particles and thrown into the blood stream to be carried to the organs of excretion. No wonder then that complications arise. To the inexperienced person it is disconcerting to find strange and new symptoms of disease developing under the Grape Cure. He needs someone with experience to explain to him that poisons which have been locked up in the system for many years have broken loose and are running riot in the blood. Hence that unusual rise of temperature, that eruption on the skin, those splitting headaches, those attacks of retching and purging, that discharge of mucus, those undue sweatings. The anxious mind of the patient should be set at rest by the assurance that all these are highly favorable symptoms of the process of purification being carried on internally—*positively prove that he is still vital enough to respond to the treatment.* The avenues of excretion—the bowels, kidneys, lungs and skin—are still in good working order. Let him then closely examine the stools, the urine, the perspiration, and let him *rejoice* with the appearance of every new evidence that Nature is still able to cast out the poisons that have been dislodged bv the magical action of the grape.

So much depends upon his mental attitude that everything should be done to enlighten him on this most important aspect of the diet.

A volume could be dedicated to the remarkable effect of the grape upon the nervous system.

Pain or discomfort is nature at work cleansing the body.

The patient should remember that every new ache and pain under the Grape Diet is an expression of life, of renewed activity. Nerves that have been atrophied for years have been stimulated by the grape.

Physical pain is Mother Nature's own voice warning us of danger. She speaks through the nerves, those delicate, watchful and intelligent protectors of the human body.

The Grape as Food and Medicine

The medicinal properties of the grape cannot be overestimated. Salts of potash are found plentifully in grapes. And now we understand why the grape may be a specific cure for cancer, for there is said to be a marked deficiency of potash in the make-up of the average cancer patient.

But it is more than that. The grape is exceptionally rich in iron and is the finest natural tonic in the world. It also has some vital relation to the protein base of the protoplasm of the cell and is on that account considered a quick re- pairer of tissue waste. As a flesh and muscle forming element it has no rival.

The grape is the perfect food, a most complete food.

Quite apart from its value as a sol- vent, it stands alone among the foods of nature as *a builder*. I have seen it ad- ministered in cases of tuberculosis and have witnessed the amazing spectacle of an emaciated patient putting on five pounds weight in one day and three pounds the next—i.e. an increase of eight pounds within forty-eight hours. This took place after a fast of six or seven days.

The grape is the most strength-giving food. It is nourishing, sustaining and completely satisfying.

Feed your typhoid patients with grapes after the acute stage and especially during convalescence, and there will probably be no relapses, no complications, no prolonged debility.

Taken without salt or sugar it relieves many complaints of the stomach, bowels and liver, because of the abundance of natural salts and acids it contains.

For Pyorrhea Poisoning

The organic acids of the grapes are strongly antiseptic and their effect on the gums is perhaps more valuable than any other result of the diet. For it means preservation of the teeth, on which mankind is dependent, not alone for health but for beauty.

Would that I had the tongue of a saint to warn against the evil of having sound teeth extracted because of poison at the roots! *It is not always necessary.* Every tooth may be loose in its socket and pus may be pouring from the gums, but after a few weeks on the exclusive grape diet it will in time be found that the teeth are firmly set in the jaws and that every trace of pyorrhea poisoning has disappeared. Do not take my word for this. *Try it. Prove it. Demonstrate the diet.*

Blood Diseases

There is said to be only one disease and that is blood disease. It is for the sake of convenience that we classify them into nervous, muscular, organic or constitutional diseases. As a matter of fact—barring accidents and malformations—we depend, for life and health, largely on the condition of the blood. And our blood is dependent, first, on what we think; second, on what we inhale; and third, on what we eat and drink.

To obtain control of these three essentials is to have perfect health.

(1) *What we think.* Since we represent the sum total of what our forefathers have thought, it were vain at this point to claim the power of independent thought. In the millennia stretching before us, maybe the Grape Cure will be found a short cut to the desired goal. I know of no way of purifying the blood —no other method as certain, sane and safe as the Grape Cure—in order to pave the way to clear thinking.

(2) *What we inhale.* Individually we have very little control over the air we

breathe. But the human race collectively is responsible for the present tainted atmosphere by which the blood is poisoned. Until provision can be made to clear the air of smoke and gas and the deadly fumes of nicotine, our bodies will suffer. The annual cleansing of the lungs by a thorough Grape Cure will be advisable.

(3) *What we eat and drink.* When all is said and done, the matter of eating and drinking is at the present time the most important because it is the only one of which we have conscious and deliberate control. We must concentrate on that, find out what are the best foods and combinations of foods and not allow ourselves to be persuaded by any one to take anything that we know to be injurious.

By right eating, it is possible to keep the blood so pure that the danger of breathing polluted air is minimized.

By right eating (and I do not think that I am going too far in saying this) we may be healthy even if we have not learned to control our thoughts. What is more, right eating will help us to obtain control over our thoughts.

To put it clearly, the condition of our blood is more dependent on the food we eat than on anything else, including thought. I know some saints whose bodies are very sick because they transgress against the laws of dietetics. And I know more than one sinner who is disgracefully healthy because he is more concerned about pure food than the good of his soul. Others there are, neither saints nor sinners, who have hardly had an original thought in their lives, and who are yet healthy and happy, like the cow placidly thinking of the next cud.

The thoughts of supermen may annul the effects of wrong eating but until we have reached that stage we do well to study the daily menu.

Purifying the Blood

I am convinced that cancer is a blood disease, more so, in fact, than any other disease. No bruise could develop into cancer if the blood were pure, and so I say again most earnestly, give the Grape Diet a trial. Go to the root of the matter and remove the cause of the trouble. Or better still prevent future trouble.

There is nothing like the Grape Diet for purifying the blood from gouty and rheumatic poisons. The inorganic deposits that have settled between the joints are apparently dissolved and expelled in the form of diarrhea or as an unpleasant oily sweat. The loosening of which may be relieved by poultices or compresses of fresh grape juice.

Pernicious anemia may be curable by a pure grape diet.

Appendicitis often loses its terrors when the secret of the grape diet has been grasped. With the grape poultice on the effected part frequently the pain subsides and the inflammation goes down.

In my opinion, the scorbutic should live entirely on grapes for a time and follow this with a diet of raw fruits, and vegetables, sour milk and cream, cheese, nuts and dried fruits.

The Quickest Relief

I used to think that fasting was the quickest relief we knew. But how much I have learned since 1925! There is no comparison between the fasting-method and the Grape Diet. I believe that nothing can take the place of the complete fast in acute disease, but the fast only partially eliminates the inorganic deposits by which chronic diseases are often caused. Perhaps that is the reason why cancer can not be cured by fasting only.

One of the great and as yet unsolved problems of medical science is, how to establish in the *diabetic*, and the *goitre* patient the normal sugar content of the blood.

A diet of raw fruits and vegetables frequently corrects these abnormal conditions of the blood. But I believe the Grape Diet is the *quickest* agent.

The marvelously rapid action of the grape must be due to the fact that grape sugar is taken immediately into the circulation without undergoing any process of digestion. There is on that account no undue tax on the organs of digestion and assimilation and Mother Nature can turn her full attention to the task of destroying the disease.

The effect is sometimes instantaneous. But again I must remind the reader that this is only true when grapes are eaten.

Unnatural Cravings

The matchless grape is the supreme remedy for the craving for alcoholic liquors. Supplying, as it does, the purest form of the alcohol, which is indispensable to the maintenance of life and health, the grape should form the exclusive diet of our unfortunate fellow creatures in the reformatories as a preliminary measure.

I am of the opinion it should be the only food used, the only nourishment permitted before and after an operation. And in our homes, every member of the family who is addicted to vice, to the drug habit and to excessive use of tobacco, tea and coffee, should be pursuaded to undergo a grape diet, without compulsion or threat. The hapless victim of perverted appetite is eager to be liberated. A sane and simple way of achieving this appeals to him.

Sex Problems

I see in the Grape Diet the solution of sex and many other social problems. By the magical purification of the blood the nerves are stabilized, self-control is established and our God-given heritage of sense and desire is transmitted into divine creative power.

No limit can be set to the stupendous importance of the Grape Cure. It is the most significant discovery of our age because the physical well-being of the entire human race may depend on it. And on this gain depends spiritual unfolding and the advancement of science.

Emancipation

This means the emancipation of the world from the iniquities of war. Let us pause a moment here—try to picture this beautiful world in a state of peace. Not a negative peace. Thrilling with life and love, supremely active, charged with the magnetic power of confidence and strength.

Simplicity

It is so simple that you can adopt it in your own home. It is only in extreme cases that patients are confined to their beds while on the Grape Diet. To be able to go about one's work as usual is perhaps one of the greatest advantages of this method. Think what it means to the business woman, the professional man, the university student! Many have taken the exclusive Grape Diet and continued their work; sometimes it is advisable to take a short exclusive diet for the week-end and after that take the following 3 stages for an equal length of time, repeating this occasionally.

Power shall be added unto power. By the purification of the blood and the general up-lifting of mind and body resulting therefrom, I see the people of earth gradually becoming more resistant to disease. Germs, plagues and epidemics lose their terror. Fearlessness (and again it is fearlessness) has been born. The entire mental outlook has been changed. Hopeful, constructive, optimistic, the sufferer sets about adjusting his faculties to this new psychology.

Appeal

Reader, are you going to do your share toward checking the devastating tide of disease and premature death that is spreading over the globe? Then do not be satisfied when, after the Grape Diet, you have regained perfect health. Your own freedom from suffering is not enough. Think of the other members of the human family and pass the message on. Tell your relatives and friends what the Grape Cure has done for you.

What we need is an

Institution

in which scientific research work can be done along these lines, for the benefit of the world. To begin with, we need a well-equipped building in which cancer patients can be treated free of charge.

We should make a public appeal for victims of this scourge to volunteer for treatment. From time to time, details should be published of the history of the patients, the methods employed in treating them, and the final results. After they are discharged, we must keep track of them, watch for a possible recurrence of the disease and treat them again if necessary.

There is no other way of bringing our testimony home to the public mind.

Who will finance so great a scheme?

It seems to me that this work should proceed from America. Fabulous sums have been spent in this country on medical research on cancer. Is it not time that natural healing had a chance?

Does the amputation of a limb or the removal of an organ remove the cause of a cancer? The cause of the trouble I contend, is still in the blood. Often a second operation becomes necessary.

The use of grapes can never restore the loss of limbs and organs, but painful swellings subside, inflammatory conditions are relieved and there is an immense relaxation from nervous strain.

Again I plead for an opportunity to *demonstrate* these wonders.

Let the medical fraternity continue their research work. It is quite possible that it may some day hit upon something more effective than the Grape Diet. Since this is more or less a cleansing diet the more you can assist nature in the elimination of these poisons - the more effective becomes the treatment itself. And above all, make certain that you have at least two or three eliminations daily. This may seem unnecessary when not eating solid foods, but you will find this to be nature's method of eliminating the poisonous wastes loosened from the tissues and various organs of the body. They must be gotten rid of - otherwise it would be like sweeping the floor over and over, without removing the dirt. The more complete the elimination the more rapid will be the results!

Chapter X

WHAT MUST WE DO?

Many people approach us with the question: "What must we do when we are threatened with an operation?" My advice would be in the first place not to let things go so far. Do not wait until it has become necessary to operate. The time to study the Grape Diet is while you are well and then the chances are that you never will be sick. The Grape Diet is so simple, that you can learn the directions by heart, so that when disease strikes your home, you may know exactly what to do. No one tries to give a drowning person a lesson in swimming. So be prepared.

But if this knowledge has come to you late, do not despair. No one in the world can force you, if you are of age, to undergo an operation. No surgeons, no medical laws can compel you to submit to the dreaded scalpel. Too often an operation is the first resort and you are rushed to the hospital in a dazed and panic-stricken state. It should be the last resort. Every other method should have been employed before you permit the delicate nerves and tissues of your body to be severed.

There is a permanent interruption in the circulation and in the flow of the vital magnetic fluid. In this book we do not dwell on the complications arising from these operations, the ruptures, adhesions, abnormal growths, weak hearts, shattered nerves and ruined digestions. Get reliable books on the subject and read—study this terrible important question from every point of view.

Ask Your Doctor

Ask your doctor to watch the results. We have met so many open-minded medical men who are fervently anxious to find a real cure for cancer, that we have no hesitation in giving this advice. Your doctor is the right man to supervise your case under the Grape Diet. He may refuse; well, then send for some other physician and if you are not successful, start right out on grapes and send for an experienced drugless physician. We want you to try your doctor first, because it is only by persuading the physicians to watch the marvelous results of this diet that we can hope to spread it over the whole world quickly.

There is one sentence which occurs in every letter written by me to the editors and physicians of this country:—

"All I ask of America, dear sirs, is an

opportunity to *demonstrate this method* in your country."

To demonstrate. Can anything be more fair, more reasonable? One who sets out to *prove* a cause must have something solid to build upon. Woe to all who set such a plea aside.

No one can heal except nature. It is by the power of mother nature that we are healed. No one can renew the oxygen in one's lungs except one's self. Ignorance of these laws of nature is keeping the world in the bondage of disease.

Your doctor may be puzzled by the strange action and reaction—we call them *healing crises*—occurring under the Grape Diet. Then ask him to consult with someone who is experienced along these lines. The Grape Diet itself is very simple but in cases of real danger the first results may be highly complicated.

Do not treat yourself without reliable advice.

Is This Faith Healing?

The laws for nature have nothing to do with faith. The sun rises on the just and unjust alike. Water quenches thirst. Food satisfies. You need no faith to enjoy them. You may bury yourself in an underground chamber and the sun will not rise on you. You may refuse to eat and drink and the laws of nature no longer exist for you. You are not co-operating with Nature.

If healing depended on faith, only the "faithful" would be well. Do we find them so? Indeed, no! There can be no more sorry spectacle on earth than the diseased, deformed, stunted, degenerated bodies of the believing (i. e. civilized) races of the earth. Compare them with the superb bodies of the heathen tribes.

I am not trying to belittle faith. Faith is the mainsping of my life, the driving-power by which I have been enabled to conquer a terrible disease. But faith alone would not have saved me. I would have died of cancer like everyone else who is afflicted with this scourge if I had depended on faith only. What can I say or do to make this point clear? *"Faith without works is dead."*

It is the want of harmony between precept and practice that we see around us that takes the heaviest toll of life. The "faithful" pray for deliverance from the consequences of their evil deeds. Prayer does not save them, because they continue the wrong habits, the sins, that have caused their diseases.

"Go and Sin No More"

Does not this divine saying imply that

disease is caused by sin? Our own or the sins of our forefathers? Transgression against the laws of nature—that is at the root of all our afflictions, and transgression, is generally due to ignorance of the law.

We appeal to our readers and fellow workers to help to combat this destructive ignorance.

The Grape Cure is a divine gift. We therefore invite our readers and patients to advance the cause by making this book known.

Educational and Preventive

Help us to check the growing danger by constructive and natural means. By surgery the evil hour is frequently postponed. In the grape diet we may have a remedy, more than that, a *sure preventative*. Join hands with us in the educational and healing campaign on which we are engaged.

This book is being published at a most auspicious moment in America. You are on the eve of the grape season. Make the coming Grape Season a memorable one by converting it into a Festival of Grapes. The news of a nation-wide healing movement in the United States will soon filter through to other countries. Great Britain will be the next to benefit, with her wide dominions. Soon the Grape Cure will be translated into the European and Oriental languages.

The Need Is Urgent

The purification of the body with a pure grape diet is going to have far-reaching results. I am not a calamity prophet, far from it, but I foresee great upheavals in the near future. There are signs in earth and sea and sky that cannot be overlooked. I use my common sense and watch the head-long course of so-called civilization. The great World War seems to have been forgotten. Something infinitely more disastrous is pending. How shall we meet it? Or better still, how can we avert it?

By the Promotion of Harmony on Earth

The old method of reforming "the other fellow" is useless. We must begin with ourselves. We must start on the lowest vehicle—the physical body—and gradually work up to perfection of mind and spirit.

Chapter XI

THE PREVENTION AND CAUSE OF CANCER

When we hear of children being born with cancer we ask, what is there in the lives of the mothers of the twentieth century that causes the transmission of this infection to their offspring? Those children have not transgressed against the laws of nature but seem to have been nourished with cancerous substances before birth.

An appalling state of affairs was revealed in an issue of the Journal of the American Association for Medico-Physical Research by the former president of the society, Dr. Frederick Dugdale, of Boston. He stated that nineteen children in every hundred who reach the age of ten show evidence of cancer.

The Use of Inorganic Substances

By watching my own case and investigating the habits of the parents of cancer patients, I have come to the conclusion that cancer is due, not so much to wrong living, artificial hours and the high pressure of modern life, as to the excessive use of common table salt, baking powder and the whole army of mineral drugs and essences prepared in the laboratory.

The human body does not possess the power of assimilating, utilizing or eliminating inorganic substances. They are of an immensely harmful nature and in times their corroding action on the nerves and tissues promotes irritating conditions in the entire system. Nature does not lie down to the task but collects the poisons and deposits them in some out-of-the-way place. A cancer is formed. That this usually happens in a weak spot, sluggish organ or bruise, is probably due to the fact that the invading substance meets with little or no resistance in tissues in which the nerves are inactive or have been injured. A bruise will not develop into a cancer in a healthy body. The cause is in the mineral-laden blood.

In South Africa, many people "draw out" cancer with herbal poultices. They have had great results in some cases. The wound caused by this process becomes a sort of safety-valve through which so much poison is eliminated that the patient is saved from an immediate recurrence of the disease. But it "gets" him in the end. So great is the suffering entailed in the "drawing" process, that many die under it.

After careful investigation of these and other methods, we come back to the grape diet. It has its limitations, like everything else, when the system of the patient is so low that there is no co-operation, but I know of no other method by which such quick results may be obtained. This is its first recommendation. There is no time to lose in treating a real cancer. The second advantage is the solvent properties of the grape. It seemingly breaks up the inorganic poisons by which cancer has been formed. Third, its powerful stimulation of the nerves under which the avenues of excretion become so magnificently active. (Note how eruptions, diarrhea and other complications occur under this drastic stimulus). And last but not least, its amazing power of building new tissue. I have said it several times but I must repeat it here. I believe the secret of the Grape Cure in wasting diseases is to be found in the rich proteids supplied by the grape.

But there must be far more than this

at the back of it and I spent many hours in deep concentration before I discovered the real explanation of the mystery.

The Most Magnetic Food

Charged with the magnetism of the sun, this Queen of Fruits, more than any other, restores and revitalizes the depleted forces of the cancer patient. Every tendril is a living receiver of cosmic magnetism. Its many-pointed leaves, forming many triangles, absorb vital essences from the air.

A perfect grape is circular in form and a bunch of grapes resembles a triangle. Students of mysticism and occultism know what these two symbols—the circle and the triangle—represent.

To come back to the cause of cancer — inorganic substances in the system. The terms *organic* and *inorganic* are much abused and little understood. Let us use the words "dead" and "alive" instead, for the sake of convenience. *Inorganic* matter is dead; *organic* matter is alive.

Man was never meant to use dead matter, such as iron, so-called refined table salt, over-cooked foods, etc., as either food or medicine. The human body does not possess the power of assimilating dead or inorganic substances. And yet it is largely composed of minerals and other "dead" matter! All foods that require cooking, should be steam-cooked.

In the first chapter of Genesis, in the wonderful allegory of the Creation, we find the explanation of this seeming inconsequence. It takes the form of dietetics and it is the only injunction laid upon man for the maintenance of life.

Behold!

"Behold, I have given you every herb bearing seed, which is upon the face of all the earth, and every tree, in which is the fruit of a tree yielding seed: *to you it shall be for meat.*"

Man was created a fruitarian. That is to say that although his body was made of the dust of the earth (minerals) his well-being depends on his use of the fruits of the earth (this term includes what we call vegetables). The first, last and only rule of life, after man had received the power of reproducing his kind and of holding dominion over living things that moved on earth and in water and air, *was concerned with food.* With an impressive "Behold!" his attention is drawn to it.

Fruit of the Vine as a Sacrament

It has taken us nearly two thousand years to understand the real significance of the introduction of the fruit of the vine as a sacrament. Is it not possible that the GREAT PHYSICIAN knew that in this delicate fruit everything was contained necessary for the healing of the body, as well as for the uplifting of the spiritual faculties?

To fashion the body of clay (minerals) and then to nourish it with vegetable matter seems ridiculous. When science discovered the elements of which the body is made and found that disease is caused by a deficiency of one or more of them, it set about the fabrication of foods, tonics, powders and pills containing these elements in an inorganic form.

Here I believe we have the cause of chronic diseases—the real cause of the increase of cancer and

That this violation of the law of nature which decrees that the animal kingdom shall feed upon the vegetable and the vegetable upon the mineral is, in the first place, the cause of cancer, and

The second cause is the use of meat— the mineral-laden blood of animals. We have to pay for every transgression and in overlooking the divine injunction to live on the fruits of the earth, we have exposed ourselves to the most ghastly, the most insidious disease.

Thirdly, *the use of cooked foods,* whether animal or vegetable.

Every kitchen stove is a laboratory on which the living essence (organic salts) is converted into dead matter (inorganic poisons.)

For years I was puzzled to know why cancer patients were unable to take cooked foods—why in my own case, relapses were brought on by the use of made dishes. I was a strict vegetarian long before I succeeded in curing myself. This proves that in a real case of cancer, it is not enough to abstain from meat. Vegetarianism alone will not save one when once the disease has taken root, not if the foods are cooked, salted or seasoned.

With the increase of cancer and tuberculosis there is a marked decrease in acute diseases.

What is the real difference between chronic and acute diseases?

Acute diseases are healing processes.

Chronic diseases are destructive processes.

Typhoid fever is evidence of Nature's efforts to expel the invading germ. A common cold is a healing process. *Cancer is not. Cancer is the death and disintegration of a given part in a living body.* I am convinced this is due to the presence of corrosive substances with which Nature is unable to cope.

Fasting does not effectually eliminate them. If the cancer is not too far advanced, a diet of raw fruits and vegetables may save the patient, but in extreme cases something is required by which those corroding, irritating substances may be quickly dissolved and expelled—something by which at the same time the strength of the patient may be nourished.

What is it in the grape that so effectually answers these three requirements? Perhaps some day science will discover the secret in the pure juice of the grape.

Chapter XII

Analysis of Grapes

Grape sugar 13.8
Tartaric acid 1.12
Nitrogenous matter .8
Gum, fat, etc. .5
Salt .... .36
Water 79.8

INSOLUBLE

Skins, stones, etc. 2.6
Pectose 9
Mineral matter .12

MINERALS PER 100 GRAMS

K_2O	11.49%	Bases	15.66%
Na_2O	0.97%	P_2O_3	7.08%
CaO	1.63%	SO_3	1.01%
MgO	1.21%	Cc	0.42%
FeO_3O_3	0.36%	Acids	8.51%

Energy per ounce 11,302

Oz. required to produce 1 oz. of protein 76.9
Oz. required to produce 1 oz. of carbo-hydr 5.2
Oz. required to produce 1 oz. of fat 62.5
Oz. required to produce 1 oz. of salt 200.0

Concord grapes contain a certain amount of grape sugar and a small percentage of iron. They are low in salts.

Green grapes are good for the complexion because they contain arsenic.

Green grapes are good for syphilis because they contain arsenic.

Grapes are numbered among the principal carbon foods.

Grapes are the third best in iron foods, containing 90%.

Grape seeds contain a great vitamin principle.

Some grapes contain 10.80% calcium, some contain a trace of iodine.

Grapes from some countries contain between 5% and 30% sugar with very little starch or protein.

Grapes from some countries contain sulphur 5.60%, magnesium 4.20%.

Grapes from some rocky countries contain boron.

Grapes yield an alkaline ash and yield alkali in the course of metabolism sufficient to change the reaction of the urine. It has been shown that this type of action is not characteristic of all fruit juices, as shown by Pickens and Hetler in "Home Economics," January, 1930.

They examined, among other things, the acidity of the urine in human subjects who had drunk large quantities of grape juice. When, under carefully controlled experimental conditions, as much as a quart of grape juice daily was ingested, neither the titratable acidity nor the hydrogen iron concentration of the urine was significantly altered. This is not the only example of the failure to decrease the acidity of the urine by a fruit with an alkaline ash.

The ash in the grape is found between the skin and fibre. It is an ash which neutralizes the toxic condition in the body, known as cancer—hence the reason for chewing and swallowing the grape skin.

(The analysis in this book was worked out by DR. FREDERICK W. COLLINS, who consulted many of the food laboratories in the world and the Department of Foods and Drugs in Washington, D. C.)

Chapter XIII

NOTHING NEW

There is nothing new about the Grape Cure. Over a hundred years ago Dr. Lambe, a pioneer reformer and dietitian, treated cancer in England with grapes. It was quite a common thing in those days to let patients loose in the vineyards of Germany, France and Italy and permit them to eat their fill. Whether other foods were included in the treatment is not known.

Even in our day the grape is well known in Europe.

Germany seems to be the centre of this natural healing cult. The grape diet is recommended by Dr. Herman Rieder, University Professor, and Dr. Martin Zeller, both of Munchen, Germany. For a complete cure these doctors prescribe the juice of freshly pressed grapes to be taken in five meals daily. Their treatment lasts from four to six weeks and the best time to undergo it is during September and October. In some cases large quantities of juice are administered —from two pounds to thirteen pounds of pressed grapes being used daily.

(We do not recommend the consumption of more than four pounds of grapes daily under most circumstances.)

In Germany, the grape is called the Queen of Fruits, and there are many well-known sanitoriums in Central Europe for the Grape Cure.

While I was experimenting on myself by fasting and dieting alternately, the facts mentioned above were not known to me. I discovered the Cure for myself in 1925, and my system of feeding the cancer patient with small quantities of grapes every two hours is altogether new. I am sure I would never have survived the terrible ordeal of that last conflict with the disease if I had not taken this natural stimulant in the form of grapes or grape juice frequently.

Its effect on the body, however, was not more wonderful than the effect it had on my mind. I had been seeking desperately for nearly nine years and now, after a seven-day fast, in satisfying a natural craving for grapes, I suddenly knew that I had struck something on which the deliverance of perishing humanity depended. A deep, inner, spiritual conviction that this simple remedy was a divine gift struck to the core of my being. Since that memorable turning-point an abiding soul-rapture has been mine. The surface storms of life can never touch it.

To share this with the world has become my highest aspiration.

The spiritual beauty of the dream grows when worked out in detail on the physical plane. Each day brings fresh surprises, new revelations, more amazing proofs. Not until you have experienced it yourself can you realize what it means to possess the power of demonstrating facts to an unbelieving world.

"In My Father's Vineyard"

Nearly 2000 years ago Someone loved to lay stress on the significance of the Vine.

His first miracle was the changing of water into wine. His last act was the abolition of the sacrifice of blood (death) and the introduction of the sacrament of wine (life.)

"I am the true Vine," He said.

We may not quite understand what

was meant by this illustration, but that is of no account. Very few of us, when we use the telephone, understand the inner workings of the device. That does not deter us from making use of it.

THE GRAPE CURE IN EUROPE

"It was very gratifying to me shortly after my arrival in America, to find that

The Grape Cure

was an old established institution. As far back as 1556 books on this wonderful Nature Cure had been published in all the various languages of Europe. On visiting a famous hospital some time ago, I learned that there is a large number of books in existence which have been published during the last four hundred years, dealing with the Grape Cure as a remedy for various diseases. The writer has been assured by an eminent authority

-151-

that this is the first book on the subject issued in America.

M. H. contributes the following:

"I was born in the vineyard districts of the Rhine Valley with its mountains on both sides and the vineyards on the foothills. The vineyards in Germany— perhaps in all Europe—are under certain and special laws for the protection of the Grapes against several kinds of insects and maladies. The farmer is not allowed to enter his own vineyard whenever he pleases, he has to arrange that certain work has to be done at times specified by the Government. Therefore sometimes the vineyard are closed by the Government and opened only to spray, etc. and for the final harvest. The Government then announces publicly (somewhere still by the Town Cryer) the fixed date for the harvest, allowing about 10-20 days for each vineyard to finish it. The help of every able bodied man, woman and child is urgently required by the Government, in order to make sure that all the Grapes are picked at the set date.—That's why all farmers, villagers and even people from nearby cities come and help with the harvest. Many are thinking of getting their yearly free Grape Cure.—And be assured, these people would never have discovered that the Grape Cure is also a "Cancer Cure," because there is no cancer amongst them, due to having their yearly Grape Cleansing.

"The vineyards are located on rocky foothills, made up terrace-like and around the plants you see only stones and pebbles covering the scant soil. The rows are planted far enough apart to permit the Spray, Fertilizer and Harvest Wagons to pass each other. At the entrance of each lane in some vineyards is a long pole which is lowered and locked at certain times by the government. When the harvesting of the Grapes is completed there is a gay and spontaneous celebration indulging in singing folksongs and dancing. From there a procession starts bringing the Grapes or Juices in to the villages in large open barrels (Bütte). In the villages the celebration continues sometimes for a week, many times the people arrayed in their National costumes.

"From the foregoing you will see, that in parts of Europe the vine industry is considered very important and therefore gets careful attention by the Government.

M. H."

Chapter XIV

DEFINITIONS GIVEN FOR THE GRAPE CURE

"GRAPE CURE—A popular method of treatment in vogue in certain parts of France, Switzerland, Germany and Tyrol, consisting of a more or less exclusive diet of grapes."

(*The Century Dictionary and Encyclopaedia*)

"GRAPE—The fruit of several species of Vitis. The cultivated grapes of the Old World are varieties of Vitis vinifera, which afford important products, namely, grapes, raisins, wine, brandy, cream of tartar, and vinegar."

(*International Encyclopaedia*)

"GRAPE CURE—A system of natural treatment in which the patient is confined wholly or chiefly to the use

-155-

of grapes for both food and drink."

(Lippincott's New Medical Dictionary)

"Pickens and Hetler examined, among other things, the acidity of the urine in human subjects who had drunk large quantities of grape juice. When, under carefully controlled experimental conditions, as much as a quart of grape juice daily was ingested, neither the iron hydrogen acidity nor the titratable concentration of the urine was significantly altered."

(Journal of the American Medical Association, Volume 94, No. 14)

"As to the curative results from X-ray and radium irradiation, these methods of treament of malignant neoplasms have proved very disappointing."

(Journal of American Medical Association

-156-

This definition is found in "A Pocket Medical Dictionary," by George M. Gould, A.M.,M.D.:

"Grape-cure.—The treatment of pulmonary tuberculosis by ingestion of quantities of grapes."

Webster has defined THE GRAPE Cure as "(Medical) Treatment of diseases, especially tuberculosis, by the free use of grapes as food."

The New International Encyclopedia, 1915, Vol. X, page 225; "A method of treatment of some diseases and conditions with a diet of which grapes form a very large part. This treatment is in vogue to some extent in France but to a much greater extent in Germany and in Hungary, in which countries as well as in Austria and Switzerland, there are sanitariums at which the Grape Cure is administered. The Grape Cure has been of advantage in diarrhea, dysentery (non-amoebic), hemorrhoids, engorgement of the spleen and has caused improvement in cases of tuberculosis, gout and some skin diseases."

GRAPELETS

When you are thirsty, drink a glass of water. When you are hungry, eat a bunch of grapes.

* * *

There is no such thing as a "fruit fast" or "grape fast." One who is eating grapes or other fruits is not fasting.

* * *

You are not on a Grape Cure when you are eating other foods with your grapes.

* * *

No one expects you to live on grapes for the rest of your life. When the grape has done its work, you go back to normal, whatever that may be.

* * *

Before you go back to it, make sure that it is not abnormal.

* * *

You do not change your religion when you go on the Grape Cure. It is simply a change of diet.

* * *

Only the sick get weak on grapes.

* * *

Never force grapes down your own or anybody else's throat. To be beneficial they must be enjoyed.

* * *

The blood is dependent on what we eat. Eat the magnetic grape and you will learn to think.

* * *

Magnetism is the connecting link between Mind and Matter. Is your link weak or strong?

* * *

No one trys to give a drowning person a lesson in swimming.

Moral—Study the Grape Cure while you are well. Be prepared.

* * *

"Feeding the patient to keep up his strength" is the surest way of killing him.

* * *

A few spoonfuls of medicine cannot undo the effects of years of wrong living.

* * *

Cast your bread upon the waters and it will come back buttered.

* * *

Acidosis is at the root of most of our bodily ailments. It is caused by the use of cooked foods, meat, starch, white bread and white sugar.

* * *

The acids in fresh fruit become sweet as honey in the stomach.

* * *

Artificial sugar, on the other hand, turns to vinegar.

* * *

Natural Healing, like all the other Laws of Nature, is an exact science.

ADDENDA

The author of this book is the first person to discover the great value of the grape and its juice *in the cure of cancer*.

Following this discovery, thousands of physicians and naturopaths throughout the world have adopted it in connection with other fruits for many chronic diseases and conditions, such as chronic constipation, hemorrhoids, chronic gastro-intestinal catarrh, hepatic ailments obesity, tuberculosis, ulcer of the duodenum, ulcer of the stomach, and cancer.

Grape vines are found growing wild throughout the temperate and parts of the torrid zones of both hemispheres. From these the cultivated varieties of the present day have been propagated. The cultivation of the grape and the making of wine are of the most remote antiquity, as appears from the scripture history of Noah and other bible characters, and for many passages of the most ancient authors, for example, Virgil and Columelia. The grape was probably introduced into the south of France and into Italy by the Phoenicians about 600 years B.C.

There are many varieties of grapes indigenous to America, particulary the Catawba, the Isabella, the Concord and California grapes, which are all described by the historian of Sir Walter Raleigh's voyages to Carolina in 1584:

During colonial times, many attempts were made to grow European varieties of grapes in the United States, but the experiment was not a success, owing to various pests and to certain mildews which did not affect the hardy vine but which infested the foreign importations. The European grapes have a higher sugar content than the American species, and are better adapted for wine making, especially champagne. They also keep better and make better raisins. The American table grapes, however, are more refreshing and make a better unfermented drink, than the imported varieties.

As a fruit, grapes are delicious, nourishing and fattening; in large quantities they are diuretic and of value in the dietetic treatment of constipation and some gastric disorders. The Grape Cure consists of eating many pounds of grapes daily. The value of grapes is due especially to their large proportion of sugar. In certain wasting diseases care may be exercised not to eat too freely, or diarrhea may be produced, which would hasten the end. Sweet grapes, when they do not purge, are, exceedingly valuable as a food and exert a curative action in bronchitis, gastric and intestinal atony. Even in Bright's disease they may, with favorable climate conditions, contribute to a restoration of health.

Finis

A TRIBUTE

It is my privilege here to offer a sincere tribute of gratitude and appreciation to those noble pioneers of Naturopathy who have done so much, suffered so much and achieved so much in America. Those who follow in their wake will find the road paved. Stumbling blocks have been removed and the forerunners will never pierce the feet of their successors. Under great tribulation, they have built up a monument of Nature Healing which will stand throughout the ages.

I advise all sufferers of serious diseases and, indeed, all who are interested in health, to study the many valuable books on Nature Cure.

To get the best permanent results, the Grape Cure should be combined with judicious fasting, deep breathing, water-treatment, sun-bathing, sea-bathing, physical exercise, fruitarianism, and the raw vegetable diet, and mind culture.

You would be amazed if you could but envisage the rapidity with which changes take place in the body while on the Grape-cure! After you have gone through the proper preparations before starting a grape-cure- in accordance with the instructions as outlined in my book   The ⟨Grape cure - the effects that you may look forward to expecting through the use of the pure, freshly crushed grape-juice, borders almost on the super-natural! Science may never discover in the laboratory the secrets of the grape, and the actual scientific causes of the results obtained, but believe me, it is far more potent than electricity! Man was created a fruitarian. That is to say that although his body was made of "dust of the earth" ie: "minerals" - his well being depends on the use of the "fruits" of the earth! (This term includes what we call "vegetables".)

Johanna Brandt